Rain

Miranda Ashwell and Andy Owen

Heinemann Library
Chicago, Illinois

Printed and bound in Hong Kong
Designed by David Oakley
Illustrations by Jeff Edwards

03 02 01 00
10 9 8 7 6 5 4 3

Library of Congress Cataloging-in-Publication Data

Owen, Andy, 1961-
Rain / Andy Owen and Miranda Ashwell.
p. cm. – (What is weather?)
Summary: Briefly discusses what rain is, how it occurs, its effects on people, plants, and animals, and the
impact of too much or too little rain.
ISBN 1-57572-789-7
Rain and rainfall–juvenile literature. [1. Rain and
Rainfall.] I. Ashwell, Miranda, 1957- . II. Title.
III. Series: Owen, Andy, 1961- What is weather?
QC924.7.O84 1999
551.57'7–dc21   98-42818
                                                                  CIP
                                                                   AC

Acknowledgments
The author and publishers are grateful to the following for permission to reproduce copyright material:
Bruce Coleman Limited/E. Bjurstrom, p. 28; M. Boulton, p. 26; J. Cancalosi, p. 27; A. Compost, p. 18;
S. Krasemann, p. 11; R. Prenzel, p. 29; H. Reinhard, p. 25; K. Rushby, p. 21; FLPA/C. Mattison, p. 19; Robert
Harding Picture Library, pp. 4, 6, 7; J. Francillon, p. 30; J. Miller, p. 9; Oxford Scientific Films/S. Olwe, p.15;
K. Wothe, p. 17; Andy Owen, p. 8; Panos Pictures/J-L Dugasi, p. 22; Planet Earth Pictures/R. Salm, p. 16;
Tony Stone Images/P. Cutler, p. 5; Still Pictures/M. Edwards, p. 20; P. Gleizes, p. 10; M. Gunther, p. 12;
R. Pfortner, p. 24; A. Watson, p. 13; Telegraph Colour Library/C. Mellor, p.14; A. Mo, p. 23.

Cover photograph: C. Brunskill/Allsport.

Some words are shown in bold, **like this**. You can find out what they mean by looking in the glossary.

Contents

What Is Rain?

Rain is water. It falls from clouds. Clouds are made of many tiny drops of water. Rain is falling from this cloud into the sea.

Rain makes everything wet. When a lot of rain falls in a short time, we say that it is a heavy rain.

Why Does It Rain?

The air is full of tiny drops of water, but they are too small to see. Sometimes there are so many drops that the air feels damp. You might even see some **mist**.

When the air cools, the water
drops join together. These larger
water drops make clouds in the sky
that you can see. The water drops
fall out of the sky as rain.

Where Does it Rain?

The wind carries clouds over high ground. The air is cold there. The cold air turns the water drops in the clouds into falling rain.

It rains when damp air cools. In **rain forests**, it rains almost every day. The warm, damp air rises. As it rises it cools, making clouds and heavy rain.

Low Clouds and Frozen Rain

Clouds sometimes lie near the ground. This is called **fog**. Driving in fog is dangerous. It is hard to see the road and other cars.

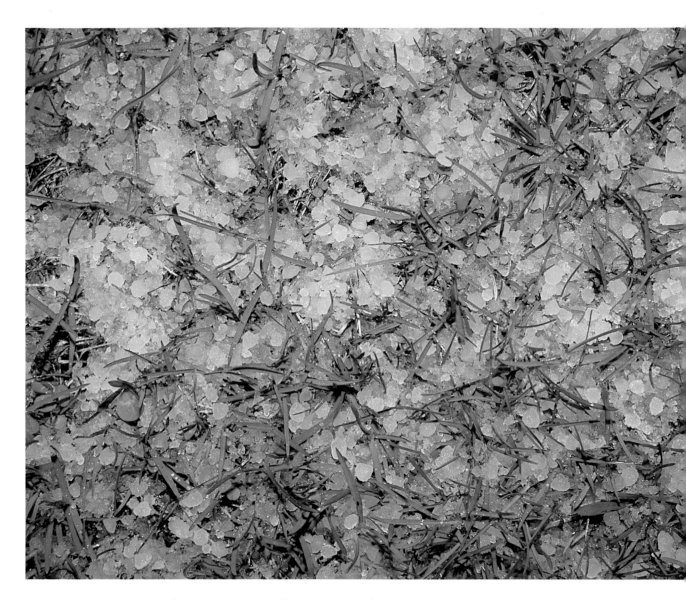

This is **hail**. Hail falls when water in clouds gets very cold. It freezes into hard pieces of ice and then falls as hail.

Wet or Dry

Rain forests have more rain than anywhere else in the world. It rains nearly every day in these hot, wet forests.

It may not rain for years in a **desert**.
Deserts can be hot or cold, but they
are always very dry places.

Rain Patterns

Rain comes at the same time every year in some places. After months without rain, the streets of cities in India are very hot and dusty.

When the rain comes, it falls very
heavily. It rains for several months.
People use the rainwater to help
grow their food.

Life in Dry Places

All animals need water to live. This camel lives where there is very little rain. It can live without water for longer than most animals. But it must drink sometime.

Plants need water to grow, or they will **wilt** and die. These plants are used to living in **deserts**. They store water inside their thick **stems**.

Life in Wet Places

Plants in the **rain forest** get plenty of rain. Rain and sunshine help the plants grow very large. They also grow very quickly.

The leaves of rain forest plants get very wet. The rain flows along the shiny leaves. Frogs live in the tiny pools of water held by the leaves.

Drinking Water

Much of the water people drink comes from rainwater. Everyone needs to drink water. Most water comes out of faucets, but some people have to collect water to take home.

When it rains, rainwater soaks into the ground. Here, a hole has been dug to reach the water. People use the water and also give some to their animals.

Collecting Rainwater

Some rainwater is held in **reservoirs**. When it has not rained for many months, there is not much water left in the reservoirs. People have used most of the water.

After weeks of rain, the reservoir is full of water. The water is carried to a city in huge pipes. It is used in homes and factories.

Dirty Rain

Smoke from factories makes the air dirty. Some of the dirt gets into water drops and falls with the rain. This is called **acid rain**.

Acid rain harms many living things.
These trees are dying because acid rain
has fallen here for many years.

Too Much Rain

Rivers flow with rainwater. After heavy rain, water can spill out of the river and onto the land. This is called a **flood**.

Floodwater flows into houses. People
can be trapped. Fast-flowing floodwater
can sweep away anything in its path.

Too Little Rain

When there is no rain, the soil dries and cracks. Most plants cannot live for long without water. They soon **wilt** and die.

A drought is when it does not rain for a long time. Animals die when they cannot find drinking water.

It's Amazing!

Sometimes we see lightning when it rains. Lightning is a huge spark of electricity that lights up the sky. Thunder rumbles as the air heats up quickly.

Glossary

acid rain	pollution mixed with rain drops
desert	place where there is very little rain
flood	when the land is covered by water
fog	clouds near the ground
hail	rain that falls as drops of ice
mist	very small water drops floating or falling near the ground
rain forest	thick forest that grows in hot, rainy places
reservoir	place where water is stored
stem	part of a plant from which the leaves and flowers grow
wilt	when plants droop and begin to dry up

More Books to Read

Davies, Kay. *Rain*. Austin, Tex: Raintree Steck-
Vaughn, 1995.

Saunders-Smith, Gail. *Rain*. Danbury, Conn:
Children's Press, 1998.

Wyler, Rose. *Raindrops and Rainbows*.
Parsipanny, NJ: Silver Burdett Press, 1989.

Index